BUILDING BLOCKS OF COMPUTER SCIENCE

# DEBUGGING

Written by Echo Elise González

Illustrated by Graham Ross

WORLD
BOOK

a Scott Fetzer company
Chicago

World Book, Inc.
180 North LaSalle Street
Suite 900
Chicago, Illinois 60601
USA

For information about other World Book publications,
visit our website at **www.worldbook.com**
or call **1-800-WORLDBK (967-5325).**
For information about sales to schools and libraries,
call 1-800-975-3250 (United States),
or 1-800-837-5365 (Canada).

Library of Congress Cataloging-in-Publication Data
for this volume has been applied for.

Building Blocks of Computer Science
ISBN: 978-0-7166-2883-5 (set, hc.)

Debugging
ISBN: 978-0-7166-2890-3 (hc.)

Also available as:
ISBN: 978-0-7166-2898-9 (e-book)

Printed in China by RR Donnelley, Guangdong Province
2nd printing August 2021

**Acknowledgments:**
Art by Graham Ross/The Bright Agency
Series reviewed by Peter Jang/Actualize
    Coding Bootcamp

# TABLE OF CONTENTS

There is a glossary on page 30. Terms defined in the glossary
are in type **that looks like this** on their first appearance.

# PESKY BUGS!

Hi! I'm a computer programming mistake, also called a computer **bug**.

Most people call me **Bug!**

Computer programmers often have to write many lines of code to create a program.

```
class CHelloWorldApplication : public CEikApplication
{
    public:
        TUid AppDllUid() const;
    protected:
        CApaDocument * CreateDocumentL();
};
```

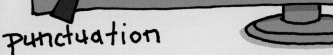

punctuation           syntax

There are many rules involved in writing the code.

With so many lines to type and so many rules to follow, it can be easy to make a mistake!

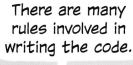

ERROR

Even the smallest mistake can result in a programming error.

Just about every programmer has had to deal with computer bugs.

Bugs are common! That's why every programmer must learn how to fix them.

Finding a **bug** can be tricky.

It can take a long time.

But finding bugs is an important step to fixing them and making a good program.

To find a bug, a programmer uses a special program called a **debugger.**

A debugger runs through the program's code line by line while the program is running.

```
8    async init ( ) {
9      const result_axis
          ={API_BASE}/setColor
          (response ⇒ {
       ...ponse.data results
13       console.log(item);
```

This process is called **stepping**.

```
8    async init ( ) {
9    const result_axis
10     get ('${API_BASE}/setColor
11     .then(response ⇒ {
12       response.data.results
13       console log(item),
```

The programmer can see each line of code as the debugger steps through them one by one.

Because the program is also running, the programmer can see the point when an error happens.

⚠ERROR

The debugger lets the programmer see which line of code is running when the error occurs.

```
response: Object
  data: Object
    results: Array
      0: Object{id: 1, price: 12}
      1: Object {id: 2, price: 20}
```

There are many different kinds of **bugs** that can occur during coding.

Some bugs occur when code does not follow the rules of the **programming language** being used.

The rules of a programming language are called **syntax**.

These pretty are flowers?

In programming, syntax includes putting words in the correct order and using the correct symbols.

These flowers are pretty!

Using the wrong word or symbol in a line of code is a common mistake.

This kind of mistake can cause a **syntax error**.

A **bug** can happen when a programmer uses a word or symbol that is not part of the **programming language** they are using.

If the word or symbol is not part of the programming language, then the computer cannot recognize it.

# TOKEN ERROR

Computer programmers call this a **token error**.

A token error is a kind of **syntax error**.

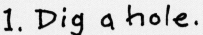

1. Dig a hole.
2. Put the roots of the plant in the hole.
3. Cover the boots with soil.
4. Water the plant.

Here are steps for planting a plant.

Can you find the bug in these steps?

3. Cover the **boots** with soil.

Step 3 contains the wrong word!

We can fix this by replacing the wrong word, *boots*, with the correct word, *roots*, in this step.

3. Cover the **roots** with soil.

Now, let's take a look at these steps for pulling a carrot out of the ground.

1. Grab the ?carrot> leaves, close to the $oil.
2. %Jiggle the carrot to l++sen it.
3. Pull the c^rrot out of ### the #soil

These steps are hard to understand!

Can you identify the bug in these steps?

Why are they so confusing?

1. Grab the ?carrot > leaves, close to the $oil.
2. %Jiggle the carrot to l++sen it
3. Pull the c^rrot out of ### the #soil

There are too many symbols that don't belong!

This is another kind of **token error.**

A computer may not understand code written using symbols that are not part of the **coding language**.

print %"Hello World"

This could result in an error.

But not to worry!

**ERROR**

1. Grab the carrot leaves, close to the soil.
2. Jiggle the carrot to loosen it.
3. Pull the carrot out of the soil.

Now that we have identified the bugs in the carrot-pulling steps, we can fix them right up.

Voilà!

1. Put on your riding boots.
2. Put on your.
3. Put the saddle on the horse.
4. Get on the horse.
5. Signal the horse to move forward .

I'm going to follow these steps to ride a horse.

---

I'm stuck on step 2! It doesn't make sense to me!

It seems like there's a word missing here...

A missing word or symbol can sometimes stop a computer from carrying out the program.

1. Put on your riding boots.
2. Put on your.
3. Put the saddle on the horse.
4. Get on your horse.
5 Signal the horse to move forward.

1. Put on your riding boots.
2. Put on your **helmet.**
3. Put the saddle on your horse.
4. Get on the horse.
5. Signal the horse to move forward.

When that happens, the programmer has to find the line of code where the word or symbol is missing and fill it in.

Giddyup!

Some **bugs** appear when the words or symbols in a line of code are out of order.

1. Bait the fishing to the attach hook.

2. Your into the water cast line.

3. Bite a fish to wait for.

4. In fish, it bites when a reel!

Umm...

These steps don't make any sense at all!

**1. Bait the fishing to the attach hook.**

Sometimes, a line of code has all the right words and symbols, but they are not in the correct order.

This can make it impossible for the computer to carry out the program.

But, the bug can be fixed...

...if the programmer puts the code elements into the correct order.

**1. Attach the bait to the fishing hook.**

**2. Cast your line into the water.**

**3. Wait for a fish to bite.**

**4. When a fish bites, reel it in!!**

# LOGIC ERRORS

The steps of a computer program also have to be arranged in the correct order.

1. Climb the ladder.

2. Pick the low apples.

3. Set a ladder against the tree.

4. Pick the higher apples.

Can you find the **bug** in these steps for picking apples?

These steps are not very efficient because they're out of order!

Each instruction is correct by itself...

But the result is not quite what we wanted!

In coding, this kind of bug is known as a **logic error**.

Unlike **syntax errors**, logic errors often do not stop the program from running.

Instead, the computer will run the program all the way through, but the result will be different than the programmer intended.

Let's fix this logic error bug by putting the steps in the correct order.

1. Pick the low apples.
2. Set a ladder against the tree.
3. Climb the ladder.
4. Pick the higher apples.

Much better!

There are other kinds of **logic errors**.

While my neighbors are away, I have to take care of their pets.

Here are the instructions my neighbor left me.

1. Sprinkle dog food in the fishbowl.
2. Walk the chickens.
3. Let the cat out of the coop.
4. Brush the fish.

I can understand these instructions.

And I can follow them.

But, they definitely don't produce the desired results!

Here, kitty kitty!!

Giving the wrong meaning to **variables** is another way a logic error can happen.

A variable is a piece of information, like a word, that can be changed in the program.

In each of these instructions, the type of pet is a variable.

1. Sprinkle <u>fish</u> food in the fish bowl.

2. Walk the <u>dog</u>.

3. Let the <u>chickens</u> out of the coop.

4. Brush the <u>cat</u>.

It looks like the wrong variables were entered!

Let's correct each of the variables...

Okay, now everyone's happy!

The best way to keep a program **bug**-free is to try to prevent bugs from happening in the first place!

There are many steps programmers can take to prevent bugs.

These steps do not prevent ALL mistakes...

But they can help to reduce the number of mistakes that happen.

It's important for programmers to test the code that they write.

Testing code can help them to see problems that might come up.

It also helps to take it slow while coding. It's easy to make a mistake when rushing!

It can also help to write your code in as simple a way as possible.

Programmers think about the simplest and most straightforward way to achieve a goal.

When code is written simply, there's less of it, and less opportunity for error.

Even those who have been coding for a long time!

It's natural to make mistakes.

With practice, you can get really good at preventing bugs...

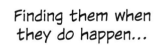

Finding them when they do happen...

And fixing them!

# GLOSSARY

**bug** an error in a program's code that keeps the program from working the way it is intended.

**coding language** (see programming language)

**debugger** a program made to find bugs in the code of another program.

**logic error** a bug that does not stop the program from working but causes it to do the wrong thing.

**programming language** a set of symbols and rules that programmers use to write computer programs.

**stepping** the process of using a debugger to check a program's code line-by-line.

**syntax** the rules that make up the "grammar" of a programming language.

**syntax error** a bug caused by improper syntax in the code.

**token error** a bug caused when code includes a word or symbol not in the programming language.

**variable** a value, or piece of information, that can change.

# GO ONLINE

Want to try out your debugging skills? Go to this website and click on the Syntax Scramble activity! Check out the other fun computer science activities and games, too!

**www.worldbook.com /BuildingBlocks**

# INDEX